T0210659

Learning Programming Using MATLAB

Learning Programming Using MATLAB
Khalid Sayood

ISBN: 978-3-031-00889-4   paperback
ISBN: 978-3-031-00889-4   paperback

ISBN: 978-3-031-02017-9   ebook
ISBN: 978-3-031-02017-9   ebook

DOI 10.1007/978-3-031-02017-9

A Publication in the Springer series
SYNTHESIS LECTURES ON ELECTRICAL ENGINEERING #3

Lecture #3
Series Editor:

First Edition
10 9 8 7 6 5 4 3 2 1

Learning Programming Using MATLAB

Khalid Sayood
Department of Electrical Engineering
University of Nebraska
Lincoln, Nebraska, USA

SYNTHESIS LECTURES ON ELECTRICAL ENGINEERING #3

ABSTRACT

This book is intended for anyone trying to learn the fundamentals of computer programming. The chapters lead the reader through the various steps required for writing a program, introducing the MATLAB® constructs in the process. MATLAB® is used to teach programming because it has a simple programming environment. It has a low initial overhead which allows the novice programmer to begin programming immediately and allows the users to easily debug their programs. This is especially useful for people who have a "mental block" about computers. Although MATLAB® is a high-level language and interactive environment that enables the user to perform computationally intensive tasks faster than with traditional programming languages such as C, C++, and Fortran, the author shows that it can also be used as a programming learning tool for novices. There are a number of exercises at the end of each chapter which should help the users become comfortable with the language.

KEYWORDS

Programming, MATLAB, Problem Solving

Contents

CHAPTER 1

Introduction

1.1 OVERVIEW

In this chapter we motivate our study of programming and attempt to justify our use of MATLAB as a tool to learn programming. We also provide a brief history of computing and suggest resources for readers interested in a more in depth treatment of MATLAB.

1.2 INTRODUCTION

Why learn programming? There are several answers to that. Computers are ubiquitous—your car, your mp3 player, the orbiting satellites which provide us with the ability to communicate and the automatic coffee maker all use a computer of some sort. And computers require programming to function. Knowing how to program provides us with a bit of insight into how our world functions. And the less mysterious our world is the more comfortable we will be in it.

Apart from the use of computers which are hidden from general view in the car or the coffee maker, depending on our particular profession, many of us will use computers directly in our professional lives. Whether we are a musician expressing ourselves through electronic compositions, an accountant doing the mysterious things accountants do, or an engineer trying to design a widget, we will end up using programs. Even if the programs you use were written by someone else, you will find when you try and use these programs for any complicated tasks you will go through a process suspiciously like programming. Albeit one which uses constructs that are specific to the profession or application.

Learning how to program is a very good way of learning how to solve problems. A program is written to solve a problem or accomplish a task. To write a successful program one has to be able to *analyze* the problem or task, and then *synthesize* the solution in the form of a program. Analysis and Synthesis are two essential aspects of problem solving. Analysis involves the breaking down of a problem into its components, while synthesis involves bring together components to make a whole. Programming initially looks like an exercise in synthesis: we put together commands and modules to perform a task. However, if we look closer, we find that programming at its heart is also an exercise in analysis. We write programs to solve problems or to achieve an objective. To understand the problem or the objective we have to first analyze it.

We have said that analysis means the breaking down of a problem into its components. The definition though is not complete without an understanding of what we mean by component. And the definition of component will vary based on context. Consider a human being. The components of the human being can be the various organs of the body, the different kinds of cells that make up the various organs, the organelles and structures that make up the cell, the various kinds of molecules that make up the structures in the cells, or the chemical elements that make up the molecules. Or, in a totally different context, the components of a human being may be the set of motivations and assumptions that govern its behavior. The set of components that will be the final product of our analysis will depend on our context. In programming the basic constructs we deal with are logical constructs. As part of learning how to program we will learn how to build logical statements and deal with the truth or falsity of logical statements. So, learning how to program provides a training in logic.

And, finally, programming is fun. It can be frustrating at times, but when you have a program that does what you want it to do it is a very satisfying. It is a creative process that exercises your brain.

Once you have analyzed the problem you are trying to solve, or the task you are attempting to accomplish, you will need to express what you want the program to do in a set of very precise instructions. Once you have a list of the precise instructions you wish to give the computer you need to translate the instructions into a language that the computer will be able to interpret. The actual instructions that a computer understands are in terms of a binary code, called the machine code, which are specific to different processors. It would be an extremely difficult task to write our instructions in binary code. Fortunately, unlike the dim dark days of yore we can write our instructions in languages that resemble English which can then be translated into something the computer can execute. These languages, called "higher level" computer languages include *FORTRAN, PASCAL, C,* and *C++.* The programs you write in these languages are translated by a program called a *compiler* into instructions the computer can act upon. First, you write a program. Then you run the compiler program (sometimes followed by a linker) which generates the instructions the computer can understand and stores them into an *executable* file. When you want to run the program you run this executable, and not the set of instructions you wrote down. There are different high-level languages that may not use a compiler to generate an executable. Instead each time you run the program the computer *interprets* your instructions, translating them into machine code, and executes them. An example of this kind of language is *BASIC,* another is *MATLAB.* As the computer has to do the translation from English-like instructions to machine code each time you run the program, programs written in these languages tend to run slower. However, the fact that the computer interprets each line can make the process of writing the program and understanding the programming process much easier. Hence, our selection of MATLAB to teach you programming. Once you understand how to program in MATLAB you will find it easy to learn other programming languages. Other reasons for introducing you to programming using MATLAB are that it is widely used in industry, many people have written programs using MATLAB that you can incorporate, and it has a very nice user interface.

1.3   ORGANIZATION AND USE

In the next chapter we spend some time looking at how to analyze a very simple problem. In the process of this analysis we describe procedures you can use when you wish to analyze especially complex tasks. The next chapter introduces you to MATLAB and gets you started. The following chapters deal with specific language aspects of MATLAB. As we work through these it is a good idea to actually implement the examples provided. It is also very important that you work through the problems at the end of each chapter. Writing a program is a very concrete activity and you can only really learn it by doing it. Therefore, doing the problems is a necessity.

1.4   WHAT THIS BOOK IS NOT

This book is not a comprehensive description of the capabilities of MATLAB. There are several very nice books out there that will provide you with a much more detailed view of MATLAB, including:

- *Introduction to MATLAB 7 for Engineers* by W.J. Palm III, McGraw Hill, 2003.
- *Mastering MATLAB 7* by D. Hanselman and B. Littlefield, Prentice Hall, 2004.

The intent of this book is to begin to teach you programming. MATLAB is only the tool we are using.

CHAPTER 2

Introduction to Programming

2.1 OVERVIEW

In this chapter we introduce you to the logic of programming and to some tools that will help you in writing programs.

2.2 INTRODUCTION

A program is a set of instructions to a computer to perform a specific task. For you to be able to write a program you need to know a *language* that the computer understands and you have to have some idea of how the computer interprets the instructions you give it. When Shakespeare has Mark Anthony say "lend me your ears," no one in the audience expects a rush to acquire sharp implements. We interpret the words in the context of our experience. This is not generally true of a computer. The computer will interpret instructions literally without making any attempt to see if the instructions are reasonable. If your instructions are not precise you will probably end up with a nonfunctional program. A computer can only do what you tell it to do. For your instructions to be precise you need to have a very clear idea of what you want to accomplish. Therefore, the first step in writing a program is analysis of the problem that the program is supposed to address.

We will begin in the next section with taking a closer look at what we mean when we talk about providing precise instructions by using what at first sight is a very simple task. We will discuss various ways in which we can take a complex problem and analyze it in order to be able to write a program to solve it. The example and its analysis might seem obvious and tedious to you, and there will be strong temptation to skip this material. However, it is very important that you spend some time with this. Hopefully, you will learn how to break down a complex problem into easily digested chunks, and how to devise a plan to achieve your objective. You will find that the time spent on devising a clear plan of attack pays for itself many times over when you begin writing the program. Finally, we will introduce you to MATLAB and begin the process of learning the language which you will use to write programs.

2.3   APPROACHING THE PROBLEM

Suppose you wanted to explain to someone who can only understand simple instructions how to compute the gas mileage for a car each time they filled up their tank. To do this you need to first figure out the information that is available to the person, and the information required from him. In more formal terms you need to know the *inputs* available, and the *outputs* required. Then you need to develop the procedure you would use to calculate the gas mileage. This step is referred to as *algorithm*[1] *development*. Finally, you need to break down the procedure into simple steps, or *refine* the algorithm so that someone who understands only very simple instructions will be able to carry out the procedure.

The reason for selecting a very simple individual as the recipient of our instructions is that in some ways the computer is very simple indeed. The computer is a very fast machine which is highly accurate and has an extremely large memory but no "understanding." It has a very limited set of instructions it "understands," and it follows these instructions exactly and literally. Hence, the need for instructions to be very precise.

The gas mileage is the number of miles traveled divided by the number of gallons of gas used. Therefore, your program needs to compute the number of miles traveled since the last fill-up, and the number of gallons used during this period. Let's suppose that you always fill up the gas tank. Therefore, the amount of gas that you put in the car is the amount used since the last fill-up. To compute the number of miles you need to subtract the odometer reading at the last time you filled up the tank from the odometer reading at this filling time. This means that you need to have saved the odometer reading from the last time you filled up. Suppose you have done so by writing the odometer reading and storing it in your glove compartment. Therefore, the inputs to your procedure are

1. The current odometer reading.

2. The amount of gas pumped.

The output of this procedure will be the mileage.

The procedure for computing the gas mileage is to retrieve the previous odometer reading and subtract it from the current odometer reading to obtain the number of miles

---

[1]The word algorithm has an interesting root. In the early ninth century Arab and Persian mathematicians were attempting to develop solutions to various linear and quadratic equations. A mathematician by the name of Al-Khwarizmi decided to abandon the idea of finding a closed form solution and instead developed a numerical approach to solving equations. Al-Khwarizmi wrote a treatise entitled *The Compendious Book on Calculation by al-jabr and al-muqabala* in which he explored (among other things) the solution of various linear and quadratic equations numerically via rules or an "algorithm." This approach became known as the method of Al-Khwarizmi. The name was changed to Algoritni in Latin, from which we get the word *algorithm*. The name of the treatise also gave us the word *Algebra*.

traveled, then read the amount of gas you pumped, and finally, divide it by the number of miles traveled.

We can write this procedure as a list of instructions:

1. Read odometer value.
2. Subtract previous odometer value from the current odometer value to obtain the number of miles traveled.
3. Divide by the number of gallons of gas pumped to determine the mileage.

This set of instructions may be sufficient for most people, but the computer needs more detailed instructions. For example, how is the computer supposed to know what the previous odometer reading was? Let's try and refine our instructions so that each step is as simple as possible:

1. Read the current odometer value.
2. Retrieve the previous value from the glove compartment.
3. Subtract the value obtained in step 2 from the value obtained in step 1.
4. Fill up the tank.
5. Read the number of gallons pumped.
6. Divide the number obtained in step 3 by the number obtained in step 5.
7. Display the number obtained in step 5 as the mileage.
8. Write the odometer value obtained in step 1 on a piece of paper.
9. Store the paper from step 8 in the glove compartment.
10. Stop.

The eighth and ninth steps are needed in order to be able to compute the mileage the next time.

Notice that each instruction is a single action. When writing a computer program you have to translate the procedure you want implemented into instructions that each consist of a single action. It is not always easy to think of a sequence of single actions that will result in a complicated procedure. However, that is how a machine works, and if you are going to "talk" to a machine you have to do so using a logic that matches the logic of the machine.

Is our set of instructions as simple as can be? Look at the last two steps. In step 8, we are trying to recall something that happened in step 1. Rather than do this we could move these last two steps up right after step 1, so that our set of instructions would read

1. Read the current odometer value.
2. Write the odometer value obtained in step 1 on a piece of paper.

3. Retrieve the previous value from the glove compartment.

4. Store the paper from step 2 in the glove compartment.

5. Subtract the value obtained in step 3 from the value obtained in step 1.

6. Fill the tank.

7. Read the number of gallons pumped.

8. Divide the number obtained in step 5 by the number obtained in step 7.

9. Display the number obtained in step 8 as the mileage.

10. Stop.

This "program" has a "bug[2]" in it. The first time we execute it there will be no paper in the glove compartment and our simple-minded friend will freak out. We wrote this set of instructions assuming that the previous odometer reading was stored in the glove compartment. This assumption will not be true the first time we use this procedure. Let's rewrite our set of instructions to fix this problem.

1. Read the current odometer value.

2. Write the odometer value obtained in step 1 on a piece of paper.

3. Is this the first time for this procedure?
   (a) If the answer is yes,

      (i) Store the paper from step 2 in the glove compartment.
      (ii) Stop.

   (b) If the answer is no, retrieve the previous value from the glove compartment.

4. Store the paper from step 2 in the glove compartment.

5. Subtract the value obtained in step 3(b) from the value obtained in step 1.

6. Fill up the tank.

7. Read the number of gallons pumped.

8. Divide the number obtained in step 5 by the number obtained in step 7.

9. Display the number obtained in step 8 as the mileage.

10. Stop.

---

[2]The word bug has been around for a long time to denote an inexplicable defect. Wikipedia has a very nice history of the use of the word.

There is a problem with this set of instructions. We have assumed that the instructions are for a simple-minded person. How can we be certain that he will remember whether he has performed this procedure before? A better way might be:

1. Read the current odometer value.
2. Write the odometer value obtained in step 1 on a piece of paper.
3. Is there a previous odometer reading in the glove compartment?
    (a) If the answer is no,
        (i) Store the paper from step 2 in the glove compartment.
        (ii) Stop.
    (b) If the answer is yes, retrieve the previous value from the glove compartment.
4. Store the paper from step 2 in the glove compartment.
5. Subtract the value obtained in step 3(b) from the value obtained in step 1.
6. Fill up the tank.
7. Read the number of gallons pumped.
8. Divide the number obtained in step 5 by the number obtained in step 7.
9. Display the number obtained in step 8 as the mileage.
10. Stop.

This way, the first time through all the person will do is write the odometer reading on a piece of paper and store it in the glove compartment.

You might think we are being a bit too picky, but you will find that being precise in the formulation of the instructions we want to give to the computer will save an enormous of time trying to fix problems caused by fuzzy statements. This precision can be something you attain by refining your instructions in stages.

2.4 FLOWCHARTS

One way to specify the necessary instructions is through an organizational tool known as a flowchart. A flowchart is a graphical way of representing a set of instructions. Each instruction is contained in a box. Different kinds of boxes are used for different statements. We will use only two types of boxes; rectangular boxes for all statements that are not questions, and diamonds for questions. The flowchart for the set of instructions shown above is shown in Figure 2.1.

It is much easier to see the "flow" of instructions with the flowchart than with the list of instructions. This also makes it easier to spot any inconsistencies in our instructions.

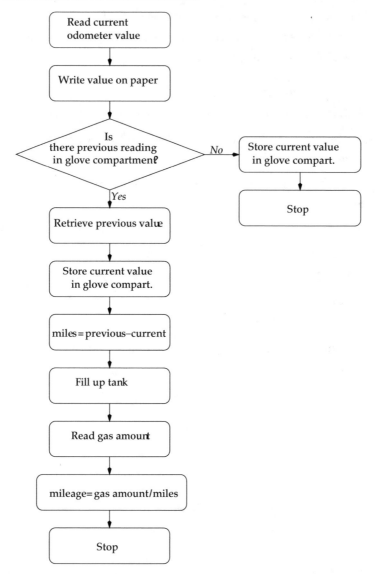

FIGURE 2.1: Flowchart for the mileage computation problem

Notice that in this case each box contains a single instruction. Once we get a flowchart with only one instruction per box we could (if we knew the language) directly translate the flowchart into a program. In this case we went directly from the analysis of the problem to a final flowchart. However, if the problem you are trying to write a program for is more complicated sometimes it is a good idea to first draw an intermediate flowchart with multiple or composite statements in a box. You can then refine the flowchart by splitting up the boxes with composite or multiple instructions into boxes containing single instructions.

Notice also that the question that we asked (both here and in our earlier list of instructions) is one that can be answered by a *yes* or a *no*. When you write a computer algorithm most of the time decisions you make during the procedure will be Yes/No or *binary* decisions. The question that you want to ask is in the form of a statement, and that statement will either be true or false.

Let's look at a variation of the mileage problem. Suppose each time you fill up your gas tank you write the odometer reading and the amount of gas on an index card. You do this many times, each time writing the odometer reading and the amount of gas on a separate card. You then stack these cards, with the most recent reading on the bottom. You want to give someone of limited intelligence a set of instructions for calculating the first ten gas mileages from the information contained on the stack of cards.

The inputs are the same as in the previous case, the outputs are ten values of mileage. The procedure would be to read the odometer values from two consecutive cards and subtract the older value from the newer value, and then divide the result by the amount of gas recorded on the card containing the newer value of the odometer reading. Let's make this procedure more precise.

Let's suppose the cards are stacked in a box labeled A. The first number on each card is the odometer reading and the second is the amount of gas pumped. For convenience, we will refer to the first number on the card we are reading as $A(1)$, and the second number as $A(2)$. To compute the mileage we need the odometer readings from two consecutive cards. Let's suppose we have another box labeled B. After we have read a card obtained from the box labeled A we will put it in the box labeled B. This is now the "previous" odometer reading for the next time we compute the mileage. As in the case of the card from the box labeled A we will refer to the first number on the card we got from the box labeled B as $B(1)$. Assuming this is not the first time through we take a card from Box A, and a card from Box B. The mileage is given by $(A(1) - B(1))/A(2)$. We can then throw away the card we got from Box B and store the card we obtained from Box A into Box B. If this is the first time through all we can do is take the card we obtained from Box A and store it in Box B.

In Figure 2.2, we show the flowchart for the procedure described above.

However, we want to compute the mileage ten times. To do this we need to keep track of how many times we have computed the mileage. Let's assume we have a tally sheet available to us, and each time we compute the mileage we make a mark on our tally sheet. We then count the number of marks we have made, and if the number of marks is less than ten we repeat the procedure. This final flowchart is shown in Figure 2.3.

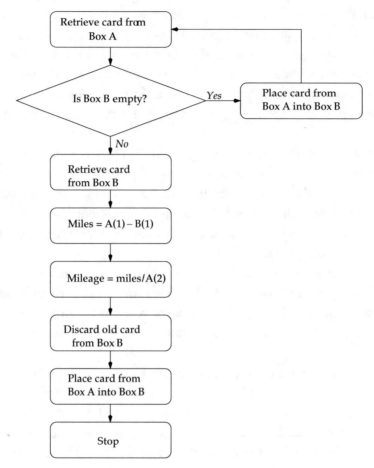

FIGURE 2.2: Flowchart for the second mileage computation problem

Once we have obtained the final flowchart or list of instructions we need to translate it into a language our simple-minded friend will understand. For us in this course, our simple-minded friend is a computer, and a language it understands is MATLAB. In the next chapter, we will begin the process of learning MATLAB and how to write instructions for MATLAB.

2.5   EXERCISES

1. Suppose you are given pairs of index cards. Each index card contains the (x, y) coordinates of two points which define a straight line. Develop an algorithm and the corresponding flowchart for determining whether these two lines intersect and, if they do the coordinates of the intersection point.

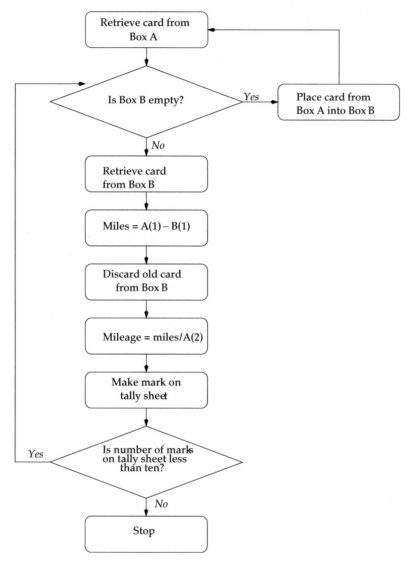

FIGURE 2.3: Final flowchart for the second mileage computation problem

2. Assume the recipient of your instruction knows how to multiply two single digit numbers. Develop the algorithm and the corresponding flowchart for multiplying two two-digit numbers

3. Generalize the algorithm to handle numbers with multiple digits. Assume you are told that one number has N digits while the other has M digits.

CHAPTER 3

Introduction to MATLAB

3.1   OVERVIEW

In this chapter we introduce some of the capabilities of MATLAB which can be used to solve problems. In the process we begin to learn how to write programs in MATLAB.

3.2   INTRODUCTION

MATLAB is a computer program originally designed to perform operations that require matrix manipulations; hence the name. Now it is a much more powerful tool and can be used to do a number of very interesting things, including solving problems which are amenable to a systematic or algorithmic approach. It has features which allow the user to manipulate speech and image data, and features which can be used to visualize all kinds of data.

Depending on the operating system you are using you can start MATLAB by either typing *matlab* in a terminal window or by clicking on the MATLAB icon. When you do this you may get a variety of windows. For now we are interested in only the "command" window. The command window will have the following prompt sign >>.

If you type help at the prompt sign you get a list of features something like this:

```
HELP topics:

matlab/general     -   General purpose commands.
matlab/ops         -   Operators and special characters.
matlab/lang        -   Programming language constructs.
matlab/elmat       -   Elementary matrices and matrix manipulation.
matlab/elfun       -   Elementary math functions.
matlab/specfun     -   Specialized math functions.
matlab/matfun      -   Matrix functions - numerical linear algebra.
matlab/datafun     -   Data analysis and Fourier transforms.
matlab/polyfun     -   Interpolation and polynomials.
matlab/funfun      -   Function functions and ODE solvers.
matlab/sparfun     -   Sparse matrices.
```

```
matlab/graph2d       -   Two dimensional graphs.
matlab/graph3d       -   Three dimensional graphs.
matlab/specgraph     -   Specialized graphs.
matlab/graphics      -   Handle Graphics.
matlab/uitools       -   Graphical user interface tools.
matlab/strfun        -   Character strings.
matlab/iofun         -   File input/output.
matlab/timefun       -   Time and dates.
matlab/datatypes     -   Data types and structures.
matlab/demos         -   Examples and demonstrations.
toolbox/local        -   Preferences.
toolbox/signal       -   Signal Processing Toolbox.
toolbox/tour         -   MATLAB Tour
```

```
For more help on directory/topic, type ''help topic.''
```

In time you may learn how to use many of these features. As we learn how to program, we will use some of them. However, before we can do that we need to become familiar with some of the more mundane aspects of MATLAB. We begin with an introduction to how numbers and characters are stored in MATLAB.

3.3   DATA REPRESENTATION

The simplest thing you can do with MATLAB is use it like a calculator. For example, if you type $4 * 5$ you get a response

```
>>4*5
ans =
    20
```

where **ans** stands for answer. What this means is that there is a location in the computer which has been tagged by MATLAB with the label **ans** and which now contains the value 20. You can assign a variable name of your choice to the result. For example, if you type $a = 4 * 5$ you get a response

```
>>a=4*5
a =
    20
```

This means that there is a location in the computer labeled by MATLAB as a which contains the value 20. If you now type $b = 5 * a$, the computer will retrieve whatever number was stored in the location labeled a (which in this case is 20), multiply that by 5 and store the result in a location labeled b.

```
>> b=5*a
b =
     100
```

Although $b = 5 * a$ looks like an equation, to MATLAB it is something else. It is an instruction to the computer to retrieve whatever is in the location labeled a, multiply it by 5 and store it in the location labeled b. This difference might become clearer if we look at the statement $b = 3 * b + 5$. As a mathematical equation this makes sense only for one value of b (-2.5). However, it means something entirely different as an instruction to the computer. As an instruction to the computer, what this says is "retrieve what was stored in b, multiply it by 3, add 5 to the result and store it back in the location labeled b." If b previously contained the value 100, it will now contain the value 305. At any time if we want to see what is stored in the location labeled b we can type b and MATLAB will respond with the contents of that particular location.

Sometimes we want to associate multiple values with a variable. For example, we want to store three test grades. If we do not want to give each grade a different name we can use an indexed array. An indexed array contains a set of values with each value being referenced by an index. For example, suppose the three grades were 76, 95, and 80. We would say

```
>>grades =[76 95 80]
grades =
   [76 95 80]
```

Now if we wished to access the second grade we could do so by typing grades(2).

We keep saying that these location contain the "value." This is not entirely true. What is contained in the location is a string of bits (0's and 1's). This can be interpreted as a binary number, or it can be interpreted as a binary code for something else. The default interpretation of the contents of a particular location is as a number. If we want to interpret the binary string as something else, we have to specify the interpretation. For example, we could interpret the contents of a particular location as an ASCII code.[1] In MATLAB you can do this by using a function called char (for character).

---

[1]The American Standard Code for Information Interchange (ASCII) is a binary eight bit code where each eight bit codeword corresponds to a printable character or a control character used for positioning of the text.

For example, if you type 22 ∗ 3 you see the following:

```
>>22*3
ans =
      66
```

However, char(22*3) will result in

```
>>char(22*3)
ans =
B
```

as 66 (or 01000010) is the ASCII code for B.

Does that mean we have to remember the ASCII code if we want to store characters at some location? Thankfully, that is not necessary. If we want the computer to interpret something as a character, or a string of characters, we simply enclose the string in single quotes.

```
>>p = 'circuit'
p =
circuit
```

From what we have seen it seems that the storage locations in MATLAB are elastic; we can store a single character or a string of characters. Actually, MATLAB assigns a sequence of locations to a specified label. These locations are organized as arrays or matrices. In fact, MATLAB was specifically designed to work with matrices. Hence, the MAT in MATLAB.

We can find out the size of the array associated with a particular label by using the size command. For example, let's look at the sizes of the arrays associated with the labels ans and p.

```
>> size(ans)
ans =
      1    1
>> size(p)
ans =
      1    7
```

Thus, ans is a label associated with storage locations organized in a 1×1 array. In other words, the label ans corresponds to a single storage location. The label p on the other hand corresponds to a 1×7 array. We will work a lot with character strings of this type which are stored in $1 \times N$ arrays. We generally will want to know the value of N and a more useful command than the `size` command for this is the `length` command which only returns the value of *N*.

```
>> length(p)
ans =
       7
```

We can examine what is contained in each of these seven locations by using an index with the label. For example, if we wished to find out what was in the third location associated with the label p, we would type p(3) and get the response

```
ans =
r
```

As expected, the third location in the array p contains the ASCII code for the character r.

You might be getting tired of seeing the `ans` statement each time we want to display something. To avoid this, we can use the MATLAB command `disp`. For example,

```
>> disp(p(3))
r
```

If you need further information about any of these commands at any time, you can always obtain it by using the `help` function. For example, suppose we wanted to obtain more information about the `char` command.

```
>> help char
```

```
CHAR Create character array (string).
   S = CHAR(X) converts the array X that contains positive integers
   representing character codes into a MATLAB character array (the first
   127 codes are ASCII). The actual characters displayed depends on the
   character set encoding for a given font. The result for any elements
   of X outside the range from 0 to 65535 is not defined (and may vary
   from platform to platform).  Use DOUBLE to convert a character array
   into its numeric codes.

   S = CHAR(C), when C is a cell array of strings, places each
   element of C into the rows of the character array S. Use CELLSTR to
   convert back.
```

```
S = CHAR(T1,T2,T3,..) forms the character array S containing the text
strings T1,T2,T3,... as rows.  Automatically pads each string with
blanks in order to form a valid matrix.  Each text parameter, Ti,
can itself be a character array.  This allows the creation of
arbitrarily large character arrays.  Empty strings are significant.

See also STRINGS, DOUBLE, STRVCAT, CELLSTR, ISCELLSTR, ISCHAR.
```

```
Overloaded methods
   help inline/char.m
   help opaque/char.m
   help sym/char.m
```

This is probably more information than you want at this time, but it is nice to know that the information is readily available when we need it.

One of the pieces of information contained in this help message is that we can use the function **double** to obtain an integer representation of the ASCII codes in the array of characters labeled p.

```
>> double(p)
ans =
      99   105   114    99   117   105   116
```

We can check our result by interpreting any of these numbers as characters. For example, let's check the character corresponding to 117.

```
disp(char(117))
u
```

Before we leave this example, notice that when we found the numeric interpretation of the ASCII codes the answer was stored in the location labeled ans and the answer contained seven elements. If we type size(ans) we will find that indeed the label ans is now associated with a 1×7 array.

Now that we have some familiarity with the functioning of MATLAB, let's see how we can write a program in MATLAB.

3.4   SCRIPT OR M-FILES

Programs in MATLAB are called *scripts* and stored in files called *M files* which have an extension of *.m*. Just as using a script allows an actor to repeat the same actions night after night, the

M-file allows you to store a sequence of actions that can be repeated by invoking the name of the file. Therefore, if you have a task that requires repetitive actions, the simplest thing to do is to create an M-file that contains those actions.

Example 3.4.1. Suppose you spend a lot of time multiplying a number by 3 and adding 5 to it. You can create an M-file (either using MATLAB's editor or your own favorite text editor) which reads as follows:

```
y = 3*x +5
```

Save this one line "program" in a file with an extension of *.m*. For example, *bob.m*. Now when we wish to execute this command we type *bob* or *run bob*. If we do not have a value stored in a location labeled x, we will get

```
>> bob
??? Undefined function or variable 'x'
>>
```

For us to get an answer, we need to define x. A simple way to do this is to set x equal to something. Suppose we set x equal to 7. Now if we type *bob* we will get

```
>> bob
y =
    26
>>
```

Why would we want to do something like this? To write a script for a single command does not make a great deal of sense. However, if there were a set of commands which we had to keep repeating all the time we might want to put them in a file. This would save us the tedium of having to type the same command over and over again. Later we will see that being relieved of this tedium allows us to write programs which require thousands, maybe millions, of repetitions of a set of instructions. This would be something we would not consider doing if we did not have the computer to do it for us. Another name given to a sequence of commands, or instructions, to the computer is *program*.

Consider the example we looked at in the last chapter where we had to set up a set of instructions for computing the gas mileage. In this example, when we are computing the gas mileage we keep repeating a sequence of actions. Let's take the situation where we have the odometer reading and the gas mileage written on individual index cards, and we repeatedly

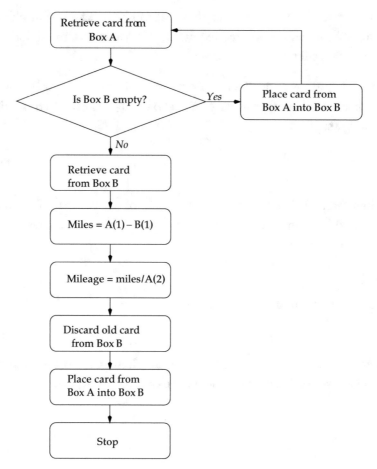

FIGURE 3.1: Flowchart for the second mileage computation problem

calculate the gas mileage. The flowchart for this situation is repeated in Figure 3.1. Let's assume that box B is not empty, and let's just write down the instructions for one pass through this process. This means we do not have to implement the instructions in the diamond-shaped boxes. We will see how to implement those particular instructions in a later chapter.

In place of box A we will define a label A with two locations, the first containing the odometer reading and the second containing the mileage. In place of box B we will use the label B to refer to an array of two locations. We write the following instructions using a text editor. An editor for creating M-files is provided with most MATLAB versions and is probably the best choice for this task. However, any text editor can be used to create the M-file. Let's suppose we wrote the following instructions and stored it in a file called *mileage.m*

```
Miles = A(1) - B(1)
Mileage = Miles/A(2)
B = A
Tally = Tally + 1
```

Before we can run this program, we need to do a couple of things to *initialize* this program. We need to provide values for the odometer readings in locations A and B, and we need to set Tally to zero. Let's do this.

```
>> B(1) = 1236
B =
        1236
>> B(2) = 12
B =
        1236          12
A = [1416  9]
A =
        1416           9
>> Tally = 0
Tally =
    0
```

Notice that we assigned the values to the locations B and A in two different ways. In the case of B, we assigned each value separately. In the case of A, we assigned both values at the same time. You can use either approach to assign a set of values to an array. Now let's run our program.

```
>> mileage
Miles =
    180
Mileage =
    20
B =
        1416           9
Tally =
    1
```

Notice that by just typing `mileage` we executed all the instructions required to compute the mileage and set the values in location B for the next cycle. At this point we can put new values for A and repeat the procedure.

```
>> A = [1584 8]
A =
          1584                8
>> mileage
Miles =
   168
Mileage =
    21
B =
          1584                8
Tally =
     2
```

Notice that the value of `Tally` has been incremented to 2. Let's run the program one more time and see what happens.

```
>> mileage
Miles =
     0
Mileage =
     0
B =
          1584                8
Tally =
     3
```

The mileage has been computed to be 0! We can easily see why this is so. At this stage the values in B are the same as in A. Therefore, when we compute the number of miles we get a value of 0. We needed to update the value of A before we ran the program. One way we can make sure that we don't make this kind of mistake is to have the computer prompt the user for the input. One way you can do this in MATLAB is to use the `input` command.

3.4.1   The Input Instruction

The format for the `input` instruction is as follows:

```
x = input('Query for the user')
```

When we run the program, the text between the quotes is printed to the screen and the computer waits for a numerical value which is then stored in the location labeled x. We can use this instruction in our mileage program by modifying the program as follows:

```
A(1) = input('What is the current odometer reading?')
A(2) = input('How many gallons of gas did you pump?')
Miles = A(1) - B(1)
Mileage = Miles/A(2)
B = A
Tally = Tally + 1
```

If we now run this program with a new odometer reading of 1703 and a gas amount of 10 gallons, we obtain the following output:

```
>> mileage
What is the current odometer reading? 1703
A =
        1703             8
How many gallons of gas did you pump? 10
A =
        1703            10
Miles =
   119
Mileage =
   11.9000
B =
        1703            10
Tally =
     4
```

where we entered the odometer reading and the amount of gas in response to prompts from the program. At this point you might notice that the program output is getting somewhat "busy." MATLAB will echo all assignments to the screen unless you suppress this feature by putting a semicolon at the end of the line. Let's do so in our program and display only the mileage value using the disp instruction. Our program now looks as follows:

```
A(1) = input('What is the current odometer reading?');
A(2) = input('How many gallons of gas did you pump?');
Miles = A(1) - B(1);
```

```
Mileage = Miles/A(2);
B = A ;
Tally = Tally + 1;
disp('The mileage is');
disp(Mileage);
```

If we now run the program with a new odometer reading of 1865 miles and 9 gallons of gas, we get the following output:

```
>> mileage
What is the current odometer reading? 1865
How many gallons of gas did you pump? 9
The mileage is
    18
```

A much cleaner output!

We have used the input function to enter numbers into our program. We can also use the input function to enter character strings with a slight modification. After we type in the prompt to the user we add , 's' to indicate to the program to expect a character string. For example,

```
A = input('What is your name?', 's')
```

will result in the program expecting a character string which will be stored in an array labeled A.

We can see this by using the following program stored in the file *nm.m*.

```
A = input('What is your name?', 's');
disp('Your name is');
disp(A);
```

If we run this program, we generate the following output:

```
>> nm
What is your name? Khalid
Your name is
Khalid
```

In this chapter we have begun our exploration of MATLAB. Before we can write useful programs in MATLAB, we will have to learn more of this language. We will continue in this endeavor in the following chapters.

3.5   EXERCISES

1. Find the *ASCII* codes for the numerals 0 through 9.

2. Find the *ASCII* codes for the lowercase alphabet *a* through *z*.

3. Find the *ASCII* codes for the uppercase alphabet *A* through *Z*.

4. What is the relationship between the ASCII codes for lowercase and uppercase letters?

5. Write a MATLAB program which will ask the user to input a number. The program should print out the square of that number.

6. Write a MATLAB program which will ask the user for their name and then write `Are you sure your name is` followed on the next line by their name.

7. Write a MATLAB program which will ask the user to enter two numbers. The program should print out the sum of the two numbers.

8. Write a MATLAB program which will take as its input an uppercase letter and by operating its *ASCII* code obtain and print out its lowercase version.

9. Write a MATLAB program which will take as its input a lowercase letter and by operating its *ASCII* code obtain and print out its uppercase version.

10. Write a MATLAB program which takes as its input a character string and print out its penultimate character.

CHAPTER 4

Selecting Between Alternatives

4.1 OVERVIEW

In this chapter we introduce structures in MATLAB which allow you to make comparisons and then execute instructions based on those comparisons. We will introduce the *if—else* structure and the *switch—case* structure. We will show how we can use these structures to compare numbers and also to compare character strings.

4.2 INTRODUCTION

In the example of the mileage computation program we did not implement the instructions contained in the diamond shaped boxes. The statements in these boxes are queries that can be answered by a *Yes* or a *No*. When a query can be answered with a *Yes* we say that the statement contained in the query is *True*. If we answer the query with a *No* we say that the statement contained in the query is *False*. We use the words *True* and *False* without any moral weight. A statement can be of the form *A is larger than or equal to B*. For different values of *A* and *B* this statement can either be true or false. We should note that when we are talking about statements of this nature, the statement is either true or it is false. It cannot be both true and false, and it cannot be neither true nor false.

4.3 COMPARING NUMBERS

When we are comparing numbers contained in two locations, say x and y, all comparisons are in terms of their position on the real number line. A number is either greater than, less than, or equal to another number. If we want to check to see if a number is negative we compare it against 0. If it is less than 0 it is negative. If not, it is positive.

Therefore, when comparing numbers the statements can be of the following forms:

$x == y$ The number in location x is equal to the number in location y
Note the double equal sign. It is very important that we have two equal signs. If we only have one equal sign MATLAB interprets this statement to be an assignment statement and copies the contents of location y into location x.

$x > y$ The number in location x is strictly greater than the number in location y. By "strictly greater than" we mean that if the numbers at the two locations are equal the statement is false.

$x >= y$ The number in location x is greater than or equal to the number in location y.

$x < y$ The number in location x is strictly less than the number in location y. Again, note the use of the word "strictly."

$x <= y$ The number in location x is less than or equal to the number in location y.

These statements are called Boolean statements as they can have only two "values" true, which corresponds to a 1, or false, which corresponds to a 0. Just as in the case of other Boolean variables we can combine two or more of these statements into a compound statement using an *AND*, *OR*, *NOT*, or *XOR*. The symbol for *AND* in MATLAB is &. The symbol for *OR* is |. The symbol for *NOT* is ~, and the symbol for exclusive or is xor. So the statement *The number in the location x is greater than the numbers in the locations y and z* is written as

$$(x > y)\&(x > z)$$

This compound statement will be true if both the simple statements that make up the compound statement are true. Similarly the statement *The number in the location x is greater than the number in location y or the number in location z* would be written as

$$(x > y)|(x > z)$$

Example 4.3.1. Let's store a value of 4 in location x and a value of 5 in location y and test the comparison statements.

```
>> x=4;
>> y=5;
>> x>y

ans =

     0
>> x<y

ans =

     1
>> x >= y
```

```
ans =
     0

>> x <= y

ans =
     1

>> x == y

ans =
     0
```

As we might expect the statements $x > y$, $x >= y$, and $x == y$, are all false and return a value of 0, while $x < y$ and $x <= y$ are both true.

Now let's store a value of 4 in a location labeled z and try out some of the compound statements.

```
>> (x>y)|(x<z)
ans =
     0
>> (x>y)|(x==z)
ans =
     1
>> (x>y)|(x>z)
ans =
     1
>> (x>y)&(x==z)
ans =
     1
```

4.4   COMPARING CHARACTER STRINGS

When we are comparing strings we need to be aware of different things then when we are comparing numbers. We might be looking for an exact match, in which case the string *adam* and the string *Adam* are different. Or, we might not care whether the letters are uppercase or lowercase as long as they match. In which case the strings *adam* and *Adam* are equal. Often we are also looking to see if the first *N* characters in two strings are the same. MATLAB provides us with functions to handle each of these eventualities. Given two strings stored in locations p and q we can use the following functions:

strcmp(p, q)   Compares the character strings in locations p and q and returns a value of true if the two strings are identical. This function is case sensitive.

strcmpi(p, q)   Similar to the strcmp function, except that it is case insensitive.

strncmp(p, q, k)   Compares only the first k characters of the strings in locations p and q and returns a true if these characters are identical.

strncmpi(p, q, k)   Similar to the strncmp function, except that it is case insensitive.

Example 4.4.1.   Let's store the character strings *adam*, *Adam*, and *Adamson* in the locations p, q, and r and test out the string compare functions.

```
>> p= 'adam';
>> q= 'Adam';
>> r= 'Adamson';
>> strcmp(p,q)

ans =

     0

>> strcmpi(p,q)

ans =

     1

>> strcmpi(p,r)

ans =

     0

>> strncmpi(p,r,4)

ans =

     1
```

As in the case of the numbers we can combine these into compound statements:

```
>> strcmp(p,q)|strncmp(q,r,4)
ans =
      1
>> strcmp(p,q)&strncmpi(p,r,4)
ans =
      1
```

4.5   IF STATEMENT

The command used in MATLAB to check whether a statement is true or not is the `if` statement. There are two different ways we can use the `if` statement. One is where if a given statement is true we want to execute a particular command. However, if it is not true we do not want to do anything. The format of such a condition is as follows:

if *statement*

    `commands to be executed if statement is true`

end

Notice the *end* statement. The *end* statement acts as a closing parentheses to the set of statements for which the *if* statement is the opening parenthesis.

    The second situation is when we want to execute one set of commands when the statement is true, and another set of commands when it is not. To do this we need one more statement which is the *else* statement. The format for the command is

if *statement*

    `commands to be executed if statement is true`

else

    `commands to execute if statement is not true`

end

For example, let's implement the function

$$y = \begin{cases} 5 \times x & x >= 10 \\ -3 \times x & x < 10 \end{cases}$$

Let's implement it in both ways described above. We could implement the two conditions $x >= 10$ and $x < 10$ separately or together. Let's first do this separately. First, of course we have to have a value for x. Let's set x equal to 3 and continue:

```
>> x=3

x =
      3

>> if x>=10
y=5*x
end
>> if x < 10
y=-3*x
end

y =
     -9

>>
```

This seems rather silly as we can see that $x < 10$ why go through the hassle of checking for it. However, if we put the conditional statements above in an *m-file*, we wouldn't even have to remember the conditions. Let's write the statements above into a file called *harry.m*. Now we can do the following:

```
>> x=3

x =
      3

>> harry

y =
     -9

>> x= 15
```

```
x =
    15

>> harry

y =
    75

>>
```

Using the else statement we can implement the function in a more compact manner as

```
if x>=10
  y=5*x
else
  y=-3*x
end
```

Note the indentations. They are not needed by the computer and the program will execute just fine without them. However, they are not just for aesthetic appeal either. They serve a very useful purpose in letting us know how the program is constructed, so if the results seem wrong we can go back and check our program. One of the things you will find out is that it is rare for a program to execute properly the first time you run it. There are always bugs that you need to remove.[1] It is a good idea to plan ahead for this eventuality and write your program so that it is easy to debug.

Example 4.5.1. Let's now try to implement the mileage program including the queries. In the flowchart, repeated in Figure 4.1, there are two decision boxes. Let's take each in turn. In the first decision the query is whether the box B is empty. As there is no physical box in the program we need to figure out a different way to ask this question. Essentially, what we want to know is if we had previously read a card from the box A, or if this is the first time. What is often done in situations like this is to use a flag. We initially set the value of the flag to 0, and then change it to 1 after whatever event we are checking for has taken place. On subsequent turns we can check the value of the flag to see if the event we are looking for has taken place or not.

---

[1]If you are in marketing you could always advertise these as features!

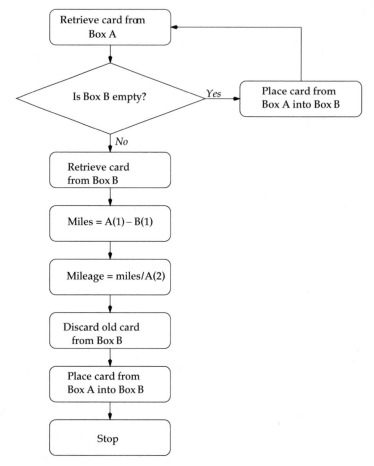

FIGURE 4.1: Flowchart for the second mileage computation problem

In the case of the second query we loop back to the beginning if the answer to the query is yes. We have not yet learned how to do loops (we will do this in the next chapter) so we will just print out a message telling us whether we have computed the mileage ten times.

```
A(1) = input('What is the current odometer reading?');
A(2) = input('How many gallons of gas did you pump?');
if flag == 0
  B = A;
  flag = 1;
else
  Miles = A(1) - B(1);
  Mileage = Miles/A(2);
```

```
  B = A ;
  Tally = Tally + 1;
  disp('The mileage is');
  disp(Mileage);
end


if Tally < 10
  disp('The number of mileage computations is less than ten')
else
  disp('We have computed the mileage ten times')
end
```

Now let's run this program:

```
>> clear all
>> Tally = 0;
>> flag = 0;
>> mileage
What is the current odometer reading? 1236
How many gallons of gas did you pump? 12
The number of mileage computations is less than ten
>> mileage
What is the current odometer reading? 1416
How many gallons of gas did you pump? 9
The mileage is
    20

The number of mileage computations is less than ten
>> mileage
What is the current odometer reading? 1584
How many gallons of gas did you pump? 8
The mileage is
    21

The number of mileage computations is less than ten
```

Notice that the first time through the program did not do any mileage computations as there were no values for *B*. Notice also that we started the process by typing the instruction clear all.

This instruction clears out all the labels so that we can start afresh. It is generally good practice to execute this instruction whenever you are starting something new.

Now let's do an example which uses the string compare function.

Example 4.5.2.   Suppose we are to write a cash register program for a store. Suppose state law requires a 10% tax on luxury items, no tax on food or medicine, and a 6% tax on everything else. Your program should query the user for the price of the item and its type, and then compute the total cost (price + tax).

```
price = input('Enter price of the item: ');
type = input('Enter type of the item: ', 's');
if strcmpi(type, 'food')|strcmpi(type, 'medicine')
  tax = 0;
else                              % if it is not food or medicine it could
  if strcmp(type, 'luxury')  % be a luxury item or not
    tax = 0.1*price;
  else
    tax = 0.06*price;
  end
end
cost = price + tax;
disp('The total cost is');
disp(cost);
```

Notice that if the item is not food or medicine it could be one of two things. To differentiate between the remaining two we have to use a second *if—else* construct nested inside the first one. Also, notice that we have some stuff written of to the side which is not really the program. MATLAB allows you to write comments on your program as long as these comments have a percent sign before them. When executing a program MATLAB ignores everything after a percent sign on a line. Commenting your program is excellent programming practice. It has two advantages. It allows you to follow your own program and helps point up any inconsistencies in logic. And, it is a great help when you come back to a program after an extended period of time.

Let's run this program and look at its output.

```
>> clear all
>> cash
Enter price of the item: 10
```

```
Enter type of the item: luxury
The total cost is
    11

>> cash
Enter price of the item: 21
Enter type of the item: food
The total cost is
    21

>> cash
Enter price of the item: 22
Enter type of the item: fuel
The total cost is
    23.3200

>> cash
Enter price of the item: 100
Enter type of the item: furniture
The total cost is
   106
```

When we have multiple possibilities we are looking for nesting *if—else* structures inside *if—else* structures can get confusing. For these situations MATLAB provides the *switch* structure. We look at this structure in the next section.

4.6   SWITCH STATEMENT

The switch statement permits us to execute different statements based on the different values of a parameter. The values can be numeric or they can be character strings.

The format of this structure is as follows:

switch *parameter*

  case *parameter value 1*

```
Statements to be executed if parameter takes on value 1
```

  case *parameter value 2*

```
      Statements to be executed if parameter takes on value 2
---

---

---

      case parameter value n

      Statements to be executed if parameter takes on value n

        otherwise

      Statements to be executed if parameter does not take on any of the
      n values.

    end
```

Let's see how this structure functions using a couple of examples.

Example 4.6.1.   Let's repeat the cash register example using the switch structure. In this case the parameter is type, and the values it can take on are food, medicine, luxury, and others. Rewriting the *cash.m* program to use the switch structure we get the following:

```
price = input('Enter price of the item:');
type = input('Enter type of the item:', 's');
switch type
  case 'food'
    tax = 0;
  case 'medicine'
    tax = 0;
  case 'luxury'
    tax = .1*price;
  otherwise
    tax = .06*price
end
cost = price + tax;
disp('The total cost is');
disp(cost);
```

If we store this program in *cash2.m* and run it we get the following output:

```
>> clear all
>> cash2
```

```
Enter price of the item: 10
Enter type of the item: luxury
The total cost is
    11

>> cash2
Enter price of the item: 2100
Enter type of the item: luxury
The total cost is
        2310

>> cash2
Enter price of the item: 25
Enter type of the item: food
The total cost is
    25
```

Because both food and medicine result in the same tax, we could have combined both of them together into a single case. We combine multiple values into a single case by enclosing the values in curly braces. If we do this our program looks as follows:

```
price = input('Enter price of the item:');
type = input('Enter type of the item:', 's');
switch type
  case {'food', 'medicine'}
    tax = 0;
  case 'luxury'
    tax = .1*price;
  otherwise
    tax = .06*price;
end
cost = price + tax;
disp('The total cost is');
disp(cost);
```

There is one problem we have not considered with this program. The *case—switch* statement is case sensitive. That is, it treats the input Food differently from the input food. To prevent the program from making an error if someone inadvertently entered Food instead of food we can use a MATLAB function lower which converts a character string to lowercase. For example,

```
>> A = lower('Food')

A =

food
```

By adding the statement

```
type = lower(type);
```

prior to the switch statement we can avoid problems that may be encountered by someone entering an input using the "wrong" case. (A counterpart to the lower function is the upper function which converts all letters in a character string to uppercase.

As we mentioned earlier the values we are looking at can be numbers as well as strings.

Example 4.6.2.   Suppose we wish to write a grading program which will take a score between 0 and 10 and provide a grade according to the following rule.

If the score is 9 or 10	the grade is A
If the score is 8	the grade is B
If the score is 5, 6, or 7	the grade is C
If the score is 4	the grade is D
If the score is below 4	the grade is F

We also want to make sure that the grade that is entered is an integer value between 0 and 10. A program which will do this is shown below:

```
x = input('Enter score on the quiz:');
switch x
  case {9,10}
   disp('Grade is A')
  case 8
   disp('Grade is B')
  case {5,6,7}
   disp ('Grade is C')
  case {4}
   disp('Grade is D')
```

```
case {0,1,2,3"
  disp('Grade is F')
  otherwise
  disp('This is not a valid score')
end
```

4.7   EXERCISES

1. Write a MATLAB program which will ask the user for a number. If the number is positive it should print out the message ⨦ The number is positive+. If the number is negative it should print out `The number is negative.`

2. Write a MATLAB program which will ask the user for two numbers (you can use the input function twice if you want). It should change the sign of the smaller of the two numbers and then add them. The program should print a message informing the user of the change in sign and also print out the final result.

3. Write a MATLAB program which asks the user to enter three numbers. The program should figure out the median value and the average value and print these out. Do not use the predefined MATLAB functions to compute the average and median values.

4. Write a MATLAB program which will ask the user to input a number. The program should print out the square of that number if the number is negative.

5. Write a MATLAB program which will ask the user for a number. If the number is equal to two times a predefined number the program should print out a welcome message. Otherwise it should print out a message saying `You are not authorized.`

6. Write a MATLAB program which will ask the user for their name. If the name matches a list of three predefined names the program should print out a welcome message. Otherwise it should print out a message saying `You are not authorized.`

7. Write a MATLAB program which will ask the user for a name and then for a password. The program should compare the name and password against a list. If both match it should print out a welcome message. If one matches it should say `Name and password do not match.` If neither matches it should say `I do not know you.`

8. Write a MATLAB program which will ask the user for a number. If the number is positive it should print out the message `The number is positive.` If the number is negative it should print out `The number is negative.`

9. Write a MATLAB program which will ask the user for their name. The program should check the name against a list of three predefined names. If it matches a name on the

list the program should print out a welcome message which is different for each name. Otherwise it should print out a message saying You are not on my list.

10. Write a MATLAB program which will ask the user for a single Roman numeral (I, V, X, L, C, or M). The program should then print out its standard Arabic equivalent. Think how you might extend this to get a Roman numeral to Arabic numeral translator.

CHAPTER 5

Loops

5.1 OVERVIEW

In this chapter we discuss different ways of making MATLAB repeat a sequence of instructions. These include the *for* loop which is used to repeat a set of instructions a specified number of times and the *while* loop which is used to repeat a set of instructions until some specified condition is met.

5.2 INTRODUCTION

We mentioned early on that one of the most useful aspects of computers was that they can efficiently perform repetitive instructions. In this chapter we look at two different methods available for performing repetitive tasks; the *for* loop and the *while* loop.

There are situations where we know ahead of time how many times a set of instructions is to be repeated. For example, if we are trying to compile a histogram of the grades students in a class have received and we know that the maximum and minimum grades are 100 and 0 and the students can only get integer points there will be a set of instructions that will have to be repeated 101 times. In this situation we would use a *for* loop. On the other hand there are many situations where we do not know ahead of time how many repetitions will need to be performed. If we are writing a program which reads an input and operates on it and we do not know ahead of time how many inputs will be available we want the program to repeat as long as an input is available and stop when there are no more inputs. In this situations we do not know ahead of time how many times the program will repeat the operations on the input. In this situation we would use a *while* loop.

5.3 FOR LOOP

When we wish to repeat a set of instructions for a set number of times we generally use a *for* loop. The *for* loop uses a counter to keep track of the number of iterations performed. The format of a *for* loop is as follows:

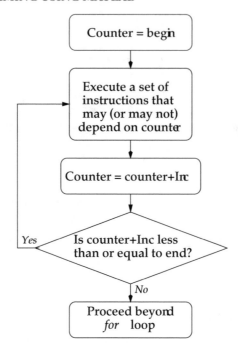

FIGURE 5.1: Flowchart for the *for* loop

for counter=begin : increment : end
    statements
    ⋮
end

begin, increment, and *end* are numbers (or variables to which we have assigned numerical values) which determine the starting and stopping values of the variable counter *counter*. The number *increment* determines how counter is incremented.

    If we were to draw a flowchart of how the for loop works it would look something like the flowchart shown in Figure 5.1.

    Let's look at an example to see how this works. Let's create an *m-file* called *becky.m* with the following content:

```
for i=0:1:5
  r = 2*i;
  p = 2*r
end
```

In this program *i* is the *counter*. The value of *begin, increment,* and *end* are 0, 1, and 5 respectively. The statements to be executed are r=2*i and p=2*r. We have not put a semicolon after the second statement so MATLAB will echo the value stored in location p to the screen each time it is changed. Let's work through this program for a few steps "by hand" before we run it. The first time through i is 0. Therefore, r will be 0 and p will be 0. Then i gets incremented by the increment value of 1. Thus the new value at location i is 1. As r=2*i the value at location r becomes 2. And, as p=2*r, the value at location p becomes 4. As the value of i is less than the *end* value of 5 we repeat the procedure incrementing the value of i by 1 to give us a value of 2. This time r=2*i becomes 4, therefore, p=2*r becomes 8. This process continues until i reaches the value of 5 generating a value of 10 for the location labeled r and 20 for the location labeled p. If we run this program we should be able to see the sequence of values at location p as i goes from 0 to 5.

```
>> becky


p =

     0
p =

     4
p =

     8
p =

    12
p =

    16
p =

    20
>>
```

If we change the increment value to 2, i will take on the values 0, 2, and 4 to give values of p of 0, 8, and 16. Note that i will not take on the value of 6 as that would be greater than 5. The program looks as follows:

```
for i=0:2:5
   r = 2*i;
   p = 2*r
end
```

Executing the program we obtain the following sequence of values.

```
>> becky

p =

     0
p =

     8
p =

    16
>>
```

While this is a good way of repeating a set of instructions, sometimes we need to stop before the counter reaches the *end* value. To do this we use the *break* statement. Let's modify *becky.m* so that we break out of the for loop whenever p gets to be greater than 3.

```
for i=0:2:5
    r = 2*i;
    p = 2*r
    if p > 3
       break
    end
end
```

Now running the program:

```
>> becky

p =

     0
p =

     8
>>
```

As we can see as soon as the value of p got to be greater than 3, execution of the instructions in the loop was terminated.

Example 5.3.1. We now have all the tools we need to implement the mileage program in its entirety. We will use the for loop to repeat the mileage calculations ten times.

```
B(1) = input('What is the current odometer reading?');
B(2) = input('How many gallons of gas did you pump?');
```

```
for tally = 1:1:10

  A(1) = input('What is the current odometer reading?');
  A(2) = input('How many gallons of gas did you pump?');
  Miles = A(1) - B(1);
  Mileage = Miles/A(2);
  B = A ;
  disp('The mileage is');
  disp(Mileage);

end
```

Notice that the first time we read the odometer reading and the number of gallons directly into the location labeled *B*. Notice also that we do not have to check to see if the tally exceeded 10. The for loop structure automatically guarantees us that the instructions in the loop will not be executed after tally reaches a value of 10.

If we run this program we get the following output:

```
>> clear all
>> mileage
What is the current odometer reading? 1236
How many gallons of gas did you pump? 12
What is the current odometer reading? 1416
How many gallons of gas did you pump? 9
The mileage is
    20

What is the current odometer reading? 1584
How many gallons of gas did you pump? 8
The mileage is
    21

What is the current odometer reading? 1826
How many gallons of gas did you pump? 11
The mileage is
    22

What is the current odometer reading? 2006
```

```
How many gallons of gas did you pump? 9
The mileage is
    20
```

```
What is the current odometer reading?
```

and the program keeps querying until the mileage has been computed ten times.

In this example the instructions within the for loop did not use the counter. Let's look at an example where the instructions being repeated are functions of the counter.

Example 5.3.2. Let's write a program which will total up all numbers between 1 and N where N is a number specified by the user. If we wrote this operation out mathematically, it would look like this

$$total = 1 + 2 + 3 + 4 + 5 + \cdots + N$$

In practice what we do is we add the first number to the total. Then we add to the total the second number which is one greater than the first number. To this total we add the third number which is one greater than the second number until we reach the Nth number. So our repetitive action is

$$total = total + i$$

where i increases by one every time. In terms of the for loop this procedure becomes

```
for i=1:1:N
   total = total +i;
end
disp(total)
```

Suppose we save this in a file called tot.m and run it.

```
>> tot
??? Undefined function or variable 'N'.

Error in ==> /home/ksayood/tot.m
On line 1  ==> for i=1:1:N
```

We never defined what N is. Let's add an instruction at the beginning to request a value for N.

```
N = input('Enter how many numbers you wish to total:');
for i=1:1:N
   total = total +i;
end
disp(total)
```

If we run this program we get the following:

```
>> tot
Enter how many numbers you wish to total: 15
Warning: Reference to uninitialized variable total in tot at line 3.
> In /home/ksayood/tot.m at line 3
```

We never initialized the value of `total`. Let's do so.

```
N = input('Enter how many numbers you wish to total:');
total = 0;
for i=1:1:N
   total = total +i;
end
disp(total)
```

This time when we run the program we get an answer.

```
>> tot
Enter how many numbers you wish to total: 15
   120
```

The reason we went through this example in this way was to show you that there are some common mistakes that you should look out for in even the simplest of programs. Not initializing variables is one of the commonest of mistakes. In this particular case we got off easy. The variable `tot` had never been used and we got a warning. What if the variable `tot` had been used previously and contained a value other than 0? In that case we would not have received a warning, our answer would have been wrong and we might not have been aware of the fact that we were operating with wrong answers. *Always initialize variables.* To prevent a variable from inadvertently using a value from a previous incarnation it is a good idea to *clear* all previous definitions. We can do this by putting the statement `clear all` as the first statement in all our programs. Our final program looks as follows:

```
clear all
N = input('Enter how many numbers you wish to total:');
total = 0;
for i=1:1:N
   total = total +i;
end
disp(total)
```

5.4   WHILE LOOPS

The *for* loop is useful when we know how many times we are going to be repeating something. However, there are many cases where we do not know ahead of time how many times we want to go through a loop. For example, suppose we wanted to write a program which would total up the purchases (along with applicable taxes) in a store. While we have to repeatedly perform the same calculations, such as computing taxes, we cannot know ahead of time how many items a particular customer may buy.

For these kinds of situations the *while* loop comes in very handy. The format of the while structure is as follows:

while *condition*   `Statements to be executed end`

A flowchart of the while loop is shown in Figure 5.2.

Example 5.4.1.   Let's write a program for the cash register. The program should ask for the price of the item and the category. If the item is either food or medicine no taxes should be added. If the item is a luxury item a tax of 10% should be added. All other items should be assessed a 6% tax. We need to find a way to signal to the program that the last item has been entered. A simple signal in this situation could be a negative or zero value for the price.

```
clear all
total = 0;
price = input('Enter price of the item:');
while price > 0
   type = input('Enter type of the item:','s');
   switch type
     case {'food','medicine'}
       tax = 0;
```

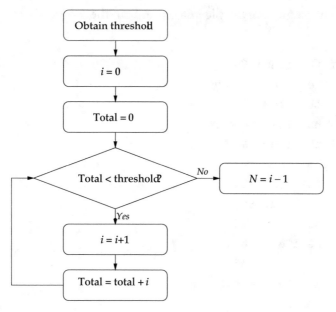

FIGURE 5.2: Flowchart for the *while* loop

```
     case 'luxury'
       tax = .1*price;
       otherwise
       tax = .06*price;
     end
     cost = price + tax;
     total = total + cost;
     price = input('Enter price of the item:');
   end
   disp('The final total is');
   disp(total);
```

Notice that we initialized `total` at the beginning of the program. Also, notice that we have two statements for reading in the price, one outside the while loop and one inside. If we did not have the input statement outside the while loop, the loop would never get started. The first time the loop tried to check the condition on the price it would not be able to do so because we do not have the price defined. Therefore the first input statement is to get the while loop started. The second query (`price=input('Enter the price of the item:'):`) inside the while loop is the one that gets repeated.

We stored this program in the file *cash3.m* and ran the program

```
>> cash3
Enter price of the item: 35
Enter type of the item: luxury
Enter price of the item: 27
Enter type of the item: food
Enter price of the item: 100
Enter type of the item: auto
Enter price of the item: 22
Enter type of the item: toy
Enter price of the item: 15
Enter type of the item: food
Enter price of the item: 56
Enter type of the item: luxury
Enter price of the item: 0
The final total is
   271.4200
```

In this example we used the value of 0 for price as the signal that we had run out of items. You have to be a bit careful with this making sure that the value you use to indicate the end of a sequence of values is not a valid value in the context of the program. Is it possible that the price of a particular item is 0? If so we could have used a negative number as our indicator. Is it possible that the price can be negative? The answer might be yes if we are including returns. In this case we might use a very large negative number which would not occur in practice, or we could manage the while loop by looking at the category, for example, stopping the loop when the category was end.

Example 5.4.2.   Let's try another version of the total problem. Instead of finding the sum from 1 to N, let's write a program that will find the largest value of N for which the total is less than some user specified threshold value. Let's begin by obtaining a flowchart for this problem. The flowchart is shown in Figure 5.3. Notice that in the end we pick N to be one less than i. The reason is that the last value of i made the total go above the threshold. Based on this flowchart we can write the program as follows:

```
clear all
Threshold = input('Enter the threshold value:');
Total = 0;
i = 0;
```

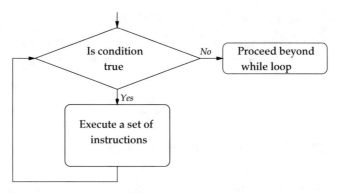

FIGURE 5.3: Flowchart for the total problem

```
while Total < Threshold
  i = i +1;
  Total = Total + i;
end
N = i - 1;
disp(N);
```

Running this program we obtain:

```
Enter the threshold value: 4097
    90
```

We can verify this result by using the program *tot*

```
>> clear all
>> tot
Enter how many numbers you wish to total: 90
     4095
```

Clearly, if we added more than 90 numbers we would get a total greater than 4097.

5.5   EXERCISES

1. Write a MATLAB program using the *for* construct which will ask the user to input a number. The program should print out the sum of all numbers from 1 to that number. For example, if the user inputs the number 10. The output should be

$$\sum_{k=1}^{10} k$$

2. Repeat the previous program using the *while* construct.

3. Write a MATLAB program which will ask the user for two numbers k and l. Using the *for* loop find the sum of all numbers between k and l, that is,

$$\sum_{j=k}^{l} j$$

4. Repeat the previous program using the *while* loop.

5. Write a MATLAB program which will ask the user for two numbers k and l. Using the *for* loop find the sum of the squares of all numbers between k and l, that is,

$$\sum_{j=k}^{l} j^2$$

6. Repeat the previous program using the *while* loop.

7. Let's build a compound interest calculator. Given the initial value X and the annual interest rate R and the maximum number of years N use a *for* loop to calculate and print out the accumulated amount for each year.

8. Suppose you are given the initial amount X and the annual interest rate R and a desired amount D. Use the *while* loop to find out the smallest number of years the money X has to remain invested at rate R for the total amount to reach or exceed D. Print out the amount accumulated after each year.

9. The following sequence is called a Fibonacci sequence:

$$1, \ 1, \ 2, \ 3, \ 5, \ 8, \ 13, \ 21, \ 34, \ 55, \dots$$

After the first two elements, each element of the sequence is the sum of the previous two elements. Write a MATLAB program which, given the first two elements, will generate and print out the next a elements of the Fibonacci sequence, where a is a number supplied by the user.

10. Write a program which will prompt the user for a predetermined word. If the word is not correct it will ask again, and will keep asking until the user enters the correct word. The program should then print out the number of tries used to guess the word.

11. Write a program which will ask the user to set up a new password. The password should be at least six characters long. If the password entered by the user is less than six characters long the program should issue a request to try again.

12. Write a program which will ask the user to set up a new password. The password should be at least six characters long and contain at least one number. If the password

entered does not satisfy these requirements the user should be prompted to try again. The process should continue until a valid password is acquired or the user gives up in disgust.

13. Write a MATLAB program for translating a number using roman numerals to the Arabic number system. For example, if the user enters VII, the program should print out 7, and if the user inputs IV, the program should print out 4. To do this it would be useful to first think about how Roman numbers are interpreted then write out your programming approach.

CHAPTER 6

Input and Output

6.1 OVERVIEW

In this chapter we introduce the different ways we can provide input to programs in MATLAB and obtain output from programs. In particular, we will look at how we can read from and write to files from a program in MATLAB.

6.2 INTRODUCTION

In the previous chapters, when we wrote programs we used the *input* and *disp* commands to obtain data for a program or to convey the results of a program. Unfortunately, these commands only allow us to interact with the program from the keyboard or the screen. Furthermore, these commands do not give us much control over how the output from a program is displayed on the screen. In this chapter we will look at ways in which we can read from, or write to, a file, with some degree of control over the form of the output. Before we can do any reading or writing we need to open the file. We see how to do this in the next section.

6.3 OPENING A FILE

Often we want to write programs which will operate on data residing in a file and write out the results of our operations to a file. To do that we need to know how to open a file and the various commands to read to, and write from, a file. By opening a file we mean verifying its existence, if it is a file we are going to read from, and providing the program with some way to access the file. The command to open a file in MATLAB is *fopen*. The command can have several arguments though we will mainly be concerned with two. The first argument is the name of the file we wish to open. If we are giving the name of the file directly to the command we have to enclose it in single quotes. The *fopen* command returns an integer value greater than two which becomes the identifier for the file. This identifier value is what we will use to read from, and write to, the file. For example, suppose we want to open a file called *test.dat* that already exists and resides in the directory we are currently working with. We type

```
>>handle = fopen('test.dat')
```

MATLAB will respond with

```
handle =
    3
```

The value stored in the location `handle` (3 in this case) is the identifier that the read and write commands will use to identify the file *test.dat*.

We could have previously stored the character string *test.dat* in some location with some label. Suppose we obtained it in response to the following command:

```
>>name = input('Enter file name','s')
```

In this case we can open the file by referring to the location that contains its name.

```
>>handle = fopen(name)
```

Notice that we did not need quotes in this case.

The second argument to the *fopen* command tells MATLAB how we wish to interact with the file. We have a number of options. We may just want to read from the file or only write to the file or, maybe do both. It is important that we make our intentions explicit for several reasons. If the file we are opening is a file that we only wish to write to, MATLAB will create a file with the name we gave if it does not already exist. If we want to read from a file MATLAB will check if it exists. If the file does not exist the *fopen* command will return a value of −1. This way we can put a check in your program to confirm that the file actually exists before we try to read from it. The *fopen* command can also return a message to us but usually it is easier to simply check the value of the identifier. We inform MATLAB about the type of interaction by placing as the second argument one of the following:

'r'   if you want to open the file only for reading
'w'   if you want to open or create a file for writing
'r+' if you want to open a file for both reading and writing
'a'   if you want to append the output of your program to a file. In this case if the file does not already exist MATLAB will create the file. If it already exists MATLAB will append the output of the program to the contents of the file.

Thus, in order to open a file for reading we would say

```
id = fopen(name,'r')
```

There are a number of other options available for the fopen command. You can get all the options by typing *help fopen*.

6.4   READING FROM A FILE

Once we have opened the file we want to read from it or write to it. Let's first look at reading from a file then we will look at how to write to a file. There are a number of ways we can read from a file. The one most often used is the command *fscanf*. The command has three arguments, the first is the file identifier obtained by using the *fopen* command. The second is the *format* in which the data in the file is written, and the third is the number of data values you wish to read. This third argument is optional. If you do not supply a third argument the entire data file will be read into an array. The second argument specifies whether we are reading a number or a string. If we are reading a number the format specifies whether we should expect digits to the right of the decimal point, and if so how many. Examples of the format specifier include:

'%d'   Read the data as integers. If you use this format statement the
        program will stop reading after the first decimal point is encountered.
'%f'   This allows for numbers with digits to the right of the decimal point.
'%s'   Reads the data into a character string. This format ignores white
        spaces and carriage control characters. If we set the third optional
        argument to 1, this format will read one space delimited word.
'%c'   This also results in the data being read into a string. However,
        white spaces and carriage control characters are preserved. If
        the third optional argument is set to 1, it reads a single character.

To see how all this works let's try out a few of these commands. We will try out these commands on two data files. The first one called *data1* looks like this

```
1 2   3   4   5
6 7 8 9 10
```

The second one called *data2* looks like this

```
3.1 4.3 5.2 3.14159
hello world
```

If you want to follow along with this example create these files using text editor **not** a word processor. The reason for this is that word processors write more than just what you type into the file. Along with what you type the word processor will include various formatting commands which allow it to determine how the file is presented to you when you open it. As far as MATLAB is concerned whatever is in the file is data so it will start reading the formatting commands as if it were data to be read. A second problem you should look out for is that the text editor might add an extension to the file name. For example, you might ask wordpad to save a file as *data1* and it might save it as *data1.txt*. Finally, throughout the following we are assuming that the file you are reading from, or writing to, is in the same location as you. To find out where you are located in the file structure type pwd in the command window. Then, either change your directory (by using cd) to where your file is, or save your files to the location where you are. To check if the file you are going to interact with is in the directory where you are located type dir in the MATLAB command window. This will also allow you to make sure that the name of the file is what you think it is.

Let's assume that the file is named correctly and is in the proper location. Let's open both these files for reading:

```
>> id1 = fopen('data1','r')

id1 =
     3

>> id2 = fopen('data2','r')

id2 =
     4
```

Now let's try and read both files using the %d format

```
>> a = fscanf(id1,'%d')

a =

     1
     2
     3
     4
     5
     6
     7
     8
     9
    10
>> b = fscanf(id2,'%d')

b =
     3
```

Notice that for data1 the read command put the contents of the entire data file into a column array. In the case of the data2 file the program read only up to the first decimal point. If we read these files using the %f format we would not see any difference in what is read from the first file however what is read from the second file changes.

```
>> frewind(id2)
>> b = fscanf(id2,'%f')


b =


    3.1000
    4.3000
    5.2000
    3.1416
```

Notice that in order to read from the beginning of the second file we needed to "rewind" it first. This time the command result in reading the numbers in the file but not the character string hello world. Furthermore, the last number has been rounded up from 3.14159 to 3.1416.

What happens if we read the files using '%s' format.

```
>> frewind(id1)
>> a = fscanf(id1,'%s')


a =


12345678910


>> frewind(id2)
>> b = fscanf(id2,'%s')


b =


3.14.35.23.1415926helloworld
```

The program reads all the characters in the file into one long string. Notice that before we could read the files we had to "rewind" them. You can think of the way the computer reads a file as how you read when you first learned to read. Most of us used our finger to mark the position of the word we were going to read. The computer does the same thing. As soon as it has read a character it moves its "finger" to the next character. If we want to reread the file we need to bring the "finger" to the beginning of the file.

We can see the size of the string we have read by using the *size* command.

```
>> size(a)

ans =

     1     11

>> size(b)

ans =

     1     28
```

Notice that all spaces in the file have been ignored. If we use the '%c' command this would not have been the case.

```
>> frewind(id1)
>> a = fscanf(id1,'%c')

a =

1 2  3   4  5
6 7 8 9 10

>> frewind(id2)
>> b = fscanf(id2,'%c')

b =
3.1  4.3  5.2  3.1415926

hello world
```

The file data2 contains both numerical values as well as a character string. If we knew this ahead of time we could read both by first using the '%f' format to read the numbers and then the '%c' format to read the character string.

```
>> frewind(id2)
>> b = fscanf(id2,'%f')
```

```
b =
     3.1000
     4.3000
     5.2000
     3.1416

>> c = fscanf(id2,'%c')

c =
hello world
```

We do not have to read the entire file in one go. By including the third argument we can dictate how many values we wish to read. For example, if we wished to read one value at a time we could put 1 as the third argument.

```
>> frewind(id2)
>> b = fscanf(id2,'%f',1)

b =
     3.1000

>> b = fscanf(id2,'%f',1)

b =
     4.3000

>> b = fscanf(id2,'%f',1)

b =
     5.2000
```

Notice that each time we read a value the program gives us the next value. Each time we read something the "finger" moves on.

Example 6.4.1. Let's rewrite our mileage program so that it reads from a file rather than from the screen.

```
clear all
name = input('Enter the name of the file containing the inputs:','s');
```

```
id = fopen(name,'r');
B = fscanf(id,'%d',2);

for tally = 1:1:10

  A = fscanf(id,'%d',2)
  Miles = A(1) - B(1);
  Mileage = Miles/A(2);
  B = A ;
  disp('The mileage is');
  disp(Mileage);

end
```

Let's save this program in the M-file *miles2.m*, and let's create a file called *miles.dat* in which we put the following data:

```
1236   12
1416    9
1584    8
1826   11
2006    9
2196   10
2323   11
2467    8
2603    8
2756    9
2918    9
```

If we run this *miles2* we get the following output:

```
>> miles2
Enter the name of the file containing the inputs: miles.dat
The mileage is
    20


The mileage is
    21
```

```
The mileage is
     22

The mileage is
     20

The mileage is
     19

The mileage is
     11.5455

The mileage is
     18

The mileage is
     17

The mileage is
     17

The mileage is
     18
```

6.5   WRITING TO A FILE

The counterpart of the fscanf command is the fprintf command. To print to a file we need to first open it. We do this using the fopen command with either the 'r+' or 'w' options. The format for the fscanf command is

```
fprintf(file_identifier, format, variable)
```

The `file_identifier` is what is returned by the fopen command. The format specifies how you want to write out the value, and `variable` is the label of the location whose contents are to be written out.

For example, if we wrote `fprintf(id,'%d',A)` the contents of location A would be written out using the %d format. The format can contain more than the placeholder %d. We could write something like `fprintf(id,'The value in A is \%d',A)`. Suppose the value in A was 5. This statement would result in the following being written to the file with identifier id

```
The value in A is 5
```

If we do not include the first parameter, the `file_identifier'` in the statement MATLAB will write out to the screen instead of a file using this format.

Example 6.5.1. Let's modify miles2.m to use the fprintf command instead of the disp command.

```
clear all
name = input('Enter the name of the file containing the inputs:','s');
id = fopen(name,'r');
B = fscanf(id,'%d',2);

for tally = 1:1:10

  A = fscanf(id,'%d',2);
  Miles = A(1) - B(1);
  Mileage = Miles/A(2);
  B = A ;
  fprintf('The mileage is %f\n', Mileage);

end
```

Notice the \n in the format. This is the newline command and it makes sure that each output is written in a new line. Notice also that we are using %f instead of %d for writing. This is because the mileage does not have to be an integer. If we run this program we get the following output.

```
>> clear all
>> miles2
Enter the name of the file containing the inputs: miles.dat
The mileage is 20.000000
The mileage is 21.000000
The mileage is 22.000000
The mileage is 20.000000
```

```
The mileage is 19.000000
The mileage is 11.545455
The mileage is 18.000000
The mileage is 17.000000
The mileage is 17.000000
The mileage is 18.000000
```

There are a lot of zeros after the decimal point. If we want to limit the number of places after the decimal point to n we would replace %f with %.nf where n is an integer. If we do this with $n = 2$ the output looks like this:

```
>> miles2
Enter the name of the file containing the inputs: miles.dat
The mileage is 20.00
The mileage is 21.00
The mileage is 22.00
The mileage is 20.00
The mileage is 19.00
The mileage is 11.55
The mileage is 18.00
The mileage is 17.00
The mileage is 17.00
The mileage is 18.00
```

This is certainly prettier looking than the previous output. In fact, formatted reads and writes give us a great deal of flexibility and allow us to create programs with natural looking outputs. Here is an example of a simple program that can be embedded into a larger program to mimic conversation.

Example 6.5.2. Let's write a program that will greet a person using their name, and a greeting that matches the time of the day. We can find the time of the day using `clock` which is a MATLAB function. We can learn more about the `clock` function using `help`.

```
>> help clock

  CLOCK  Current date and time as date vector.
     CLOCK returns a six element date vector containing the
     current time and date in decimal form:
```

```
CLOCK = [year month day hour minute seconds]
   The first five elements are integers. The seconds element
   is accurate to several digits beyond the decimal point.
   FIX(CLOCK) rounds to integer display format.

See also DATEVEC, DATENUM, NOW, ETIME, TIC, TOC, CPUTIME.
```

Therefore, if we are interested in the time of day in hours we need to look at the fourth element of this vector. By looking at the fourth element of the vector returned by `fix(clock)` we can decide whether it is morning, evening, or afternoon.

Here is a program that does this:

```
name = input('Hi, what is your name?','s');
time = fix(clock);
switch time(4)
 case{0,1,2,3,4,5,6,7,8,9,10,11}
   greet = 'Good morning';
 case{12,13,14,15,16}
   greet = 'Good afternoon';
 case{17,18,19,20,21,22,23,24}
   greet = 'Good evening';
end

fprintf('%s %s \n',greet,name);
```

If we run this program at 5:30 pm we get the following output.

```
>> hello
Hi, what is your name? Zorro
Good evening Zorro
```

We can add to this program to make it into a conversation program.

Finally, once we are done with reading from or writing to a file we need to close the file. Closing the file is done automatically in languages like *C* but this is not true for MATLAB. If you fail to close a file in MATLAB you might not be able to read it using other programs. Closing a file is easy. You use the command *fclose(id)* where *id* is the file identifier.

6.6   EXERCISES

1. Write a program which will open a file called *example.dat*. It should then ask the user for their name and age which it should write to the file. For example, if the user gives the name as *John* and the age as *18*, the program should write *John is 18 years old* to the file *example.dat*.

2. Create a file using notepad or some other text editor (**not** a word processor) and write 5 numbers in it, one number to a line. Write a program which will ask the user for the name of the file. It should then read the numbers from this file, add them up, and write the answer to the screen. For example, if the file name is *aaa.txt* and the sum of the numbers is 27, the screen message should read *The sum of the numbers in aaa.txt is 27.*

3. Create a file using notepad or some other text editor containing grades for a class. The file should look something like this:

```
Alice 10
Bob    9
Carol  7
David  9
Emily  8
End    0
```

You can have as many names as you like but the last name should be End. Now write a program which will ask you for the name of the file, read the values in the file and give you an average grade.

4. Use the file created for the previous program. Write a program which will ask you for the name of the file then ask you for the name of the student. It should then provide you the grade for the student. If the name does not appear in your roster it should put out an appropriate message.

CHAPTER 7

Functions

7.1   OVERVIEW

In this chapter we introduce the concept of functions and look at some examples of functions.

7.2   INTRODUCTION

When we write a lot of programs we will find that there are certain kinds of computations or activity we perform quite often. In these situations it is generally a good idea to write a MATLAB *function* that can be used by many programs.

Recall that a function in mathematics is a rule by which an input variable is transformed into an output variable. If we have a function

$$f(x) = x^3 + 4x^2 + 2x + 3$$

then $y = f(5)$ means: Take 5. Multiply it with itself three times. Add to the resulting sum 5 multiplied by itself then multiplied by 4. Add this to 5 multiplied by 2, and then finally add all of this to 3. Looked at this way we can see that the function notation is really a shorthand for computing an output value given an input value and a set of rules. Using the function notation we could therefore replace the problem-dependent portion of the program by a statement like

```
computed = function_name(inputs);
```

where `inputs` are the values the function would need to compute the output. In the case of the function $f(x)$ described above, the input would be the value of x.

Suppose we wrote a program to calculate the maximum value of $f(x)$. There are a number of different algorithms around for finding maxima. The structure of the algorithm itself is generally independent of particular form of $f(x)$. The only part of the algorithm that is dependent on the particular $f(x)$ being maximized is the computation of $f(x)$ for different values of x. Therefore, we could write the program in such a way that all such computations are performed using a MATLAB function. We could then use this program to find the maxima of a variety of functions. Each time we changed problems we could change `function_name` to correspond to the problem.

As you might expect you have to write down the rules by which the output of the function is generated. We describe these rules in the following section.

7.3   RULES FOR WRITING FUNCTIONS

According to the rules of MATLAB the rules that describe how the outputs are to be generated from the inputs have to be written in a file with the same name as the function name with a *.m* extension. The first statement (after any comments) is of the form

```
function[output_variables] = function_name(input_variables)
```

Notice that we can have more than one output variable and we can have more than one input variable.

Example 7.3.1. Let's take a page from Al-Khwarizmi, but let's go one better. Let's write a function that gives us the real roots of a quadratic equation $ax^2 + bx + c$, if they exist. So, the function input variables would be a, b, and c. The output variables would be the two roots and some other variable which informs the user if there are no real roots. A possible implementation of this function is as follows:

```
function [root1,root2,condition] = quadratic(a,b,c);
% This function computes the roots of a quadratic equation, if
% the equations has real roots.  In these cases the function
% returns a condition value of 0.  If the roots are complex
% the function returns a condition value of -1

%check to see if roots are real

det = b^2 - 4*a*c;
if det < 0        % Check to see if the equation has real roots
   condition = -1 % No real roots
else
   condition = 0
   root1 = (-b + sqrt(det))/(2*a);
   root2 = (-b - sqrt(det))/(2*a);
end
```

As the name of the function is *quadratic*, it should be stored in a file called *quadratic.m*. To use this function from within MATLAB in order to obtain the roots of the equation $2x^2 + 4x + 1 = 0$ we type

```
>> [x,y,z] = quadratic(2,4,1)
```

Based on how the function is written, x will contain the first root, y will contain the second root, and z will be 0 if the equation has real roots and 1 if it does not. The response from MATLAB is

```
x =
   -0.2929


y =
   -1.7071


z =
     0
```

Therefore the roots are -0.2929, and -1.7071. See if you can make *quadratic* more general by returning the real and imaginary parts of the roots as the first and second output variables when the roots are complex.

Note the comment lines after the function declaration. These lines are the help statement for this function. That is, if you type help quadratic you will get a response

```
   This function computes the roots of a quadratic equation, if
   the equations has real roots. In these cases the function
   returns a condition value of 0. If the roots are complex
   the function returns a condition value of -1
```

We can also use the MATLAB function to clean up our cash register program.

Example 7.3.2. In our cash register example we can write a function that computes the tax.

```
function[tax] = compute_tax(price,type);
%This function computes the sales tax on different items
% using the following schedule:
%                 food, medicine  - no tax
%                 luxury item     - 10% tax
%                 all other items - 6% tax
```

```
%
  switch type
    case {'food','medicine'}
      tax = 0;
    case 'luxury'
      tax = .1*price;
    otherwise
      tax = .06*price;
  end
```

We can now change our cash register program to read

```
total = 0;
price = input('Enter price of the item:');
while price > 0
  type = input('Enter type of the item:', 's');
  tax = compute_tax(price,type);
  cost = price + tax;
  total = total + cost;
  price = input('Enter price of the item:');
end
a = sprintf('The final total is %d n',total);
disp(a);
```

Notice that the new program is much more readable than the old one. Furthermore, if the tax structure is changed all we need to do is replace the compute_tax function.

While we can write our own MATLAB functions, we should be aware of various existing MATLAB functions. We list a few in the next section.

7.4   MATLAB FUNCTIONS

MATLAB contains a very large number of functions. If we type help while in MATLAB we get (among other things) a listing of these categories. They include:

elfun These are the elementary mathematical functions including the various trigonometric functions, logarithms and exponential functions, functions for manipulating complex numbers, modulus functions, and functions for rounding (up or down) numbers.

specfun These are more specialized mathematical functions such as various Bessel functions, the gamma function, and many others used in analytical treatment of systems. They also include a function for generating prime factors, another for checking if a number is prime, functions for finding the least common multiple and the greatest common divisor, and functions for converting between various coordinates.

matfun These are various matrix functions. This is an especially rich set of functions as MATLAB was originally designed for dealing with matrices. To appreciate this richness we need to have a better understanding of matrix operations.

datafun These are a large number of functions that can be useful in data analysis. There is a function for finding the maximum of a set of numbers, another for finding the minimum. Also included are functions used in statistical analysis. There are also functions for simulating filters, and functions for manipulating sound files.

polyfun These include functions for interpolation, for geometric analysis, and for differentiating and multiplying polynomials. Also included is a function for finding roots of a polynomial (much more general than our `quadratic` function.

funfun In spite of its name this group of functions has more to do with solving differential equations than with weekends in the sun. The functions in this category also include a function used for numerical integration and functions for plotting functions.

strfun This group contains functions that manipulate character strings. There are functions to create character strings, test character strings (for example, to see if a variable contains a character string), compare strings, concatenate strings, convert strings to uppercase and lowercase, and to convert numbers to strings.

iofun These are functions used for input and output. They include the familiar `fopen`, `fscanf`, and `fprintf` commands as well as functions which write to and read from strings. There are also functions for file positioning and file name handling.

timefun These include functions which return the current date and time as well as calendar functions and functions that act as stopwatches and timers.

Even though this is just a partial listing of the functions included in MATLAB we can get some idea of how many functions are already available. We can then add to these by writing our own functions making MATLAB a truly versatile language.

Let's look at some of these functions in a bit more detail.

Example 7.4.1. Let's take a look at the function for finding the maximum value of a set of numbers. To find more about this function we can use `help`.

```
>> help max
```

```
MAX    Largest component.
    For vectors, MAX(X) is the largest element in X. For matrices,
    MAX(X) is a row vector containing the maximum element from each
```

```
column. For N-D arrays, MAX(X) operates along the first
non-singleton dimension.

[Y,I] = MAX(X) returns the indices of the maximum values in vector I.
If the values along the first non-singleton dimension contain more
than one maximal element, the index of the first one is returned.

MAX(X,Y) returns an array the same size as X and Y with the
largest elements taken from X or Y.  Either one can be a scalar.

[Y,I] = MAX(X,[],DIM) operates along the dimension DIM.

When complex, the magnitude MAX(ABS(X)) is used.  NaN's are
ignored when computing the maximum.

Example: If X = [2 8 4    then max(X,[],1) is [7 8 9],
                  7 3 9]

    max(X,[],2) is [8    and max(X,5) is [5 8 5
                    9],                   7 5 9].

See also MIN, MEDIAN, MEAN, SORT.
```

Let's try our own example:

```
>> x = [7 6 9 3 1 4 8]
x =
     7    6    9    3    1    4    8
>> [y i] = max(x)
y =
     9
i =
     3
```

The max function not only returns the maximum value but also the location of the maximum
value in the array.

Example 7.4.2. Let's take a look at two string functions that you might find useful. The first, findstr finds the location of the occurrence of one string of characters in another string of characters.

```
>> help findstr

 FINDSTR Find one string within another.
    K = FINDSTR(S1,S2) returns the starting indices of any occurrences
    of the shorter of the two strings in the longer.

    Examples
        s = 'How much wood would a woodchuck chuck?';
        findstr(s, 'a')    returns  21
        findstr(s, 'wood') returns  [10 23]
        findstr(s, 'Wood') returns  []
        findstr(s, ' ')    returns  [4 9 14 20 22 32]

    See also STRCMP, STRNCMP, STRMATCH.
```

Let's suppose we have a program which inquires of the user how they are feeling. We want to check whether the user used a particular word to describe their situation. Let's define the following

```
>> word1 =  'fine'

word1 =
fine

>> word2 = 'I am fine'

word2 =
I am fine

>> word3 =  'I am ill'

word3 =
I am ill
```

where *word1* is the word we are looking for, and *word2* and *word3* are two possible responses from the user. We can use the `findstr` command to perform the search.

```
 i = findstr(word1,word2)
i =

      6
>> i = findstr(word1,word3)
i =

      []
```

In the first case, the word *fine* occurred in the second string at location 6, therefore the function returned the value of 6. In the second case, the word did not occur in the second string so the function returned a null set.

Sometimes we are simply interested in finding whether a particular word occurred in a given string, not necessarily what its location was. In these situations the Boolean function `isempty` can be very useful.

```
>> help isempty

  ISEMPTY True for empty matrix.
     ISEMPTY(X) returns 1 if X is an empty array and 0 otherwise. An
     empty array has no elements, that is prod(size(X))==0.
```

We can use these two functions as part of a dialog program in the following manner:

```
 if isempty(findstr(word1,word2))
      resp = input('Why are you not fine?', 's')
    else
      resp = input('What is it that is making you feel so good?', 's')
    end
```

7.5   EXERCISES

1. Write a MATLAB function called *larger* which takes as its input two numbers and returns the larger of the two. Write a test program to test the function.

2. Write a MATLAB function called *rom2ara* for translating a number using roman numerals to the Arabic number system. For example, if we type `rom2ara(VII)`, the function responds with 7.

3. Write a MATLAB function called *numcheck* which will take as its input a string and check to see if it contains only numeric characters. It should return a value of 1 if the input is purely numeric and zero if it is not.

4. Write a function called *alphcheck* which will take as its input a string and check to see if it contains only alphabetic characters. It should return a value of 1 if the input is purely alphabet and zero if it is not.

5. Combine the previous two functions into a MATLAB function called *incheck*. The function takes a string as its input. It checks to see if the string contains only numbers, only alphabetic characters, or neither. It should return a string naming the situation.

Printed in the United States
by Baker & Taylor Publisher Services